POLLUTION FREE ENVIRONMENT
(FROM FUELS AND OILS)
FOR THE GENERATIONS

POLLUTION FREE ENVIRONMENT
(FROM FUELS AND OILS)
FOR THE GENERATIONS

Bhupindar Singh

Copyright © 2014 by Bhupindar Singh.

| ISBN: | Softcover | 978-1-4990-0826-5 |
| | eBook | 978-1-4990-0825-8 |

All rights reserved. No part of this book may be reproduced or transmitted in any form or by any means, electronic or mechanical, including photocopying, recording, or by any information storage and retrieval system, without permission in writing from the copyright owner.

Any people depicted in stock imagery provided by Thinkstock are models, and such images are being used for illustrative purposes only.
Certain stock imagery © Thinkstock.

This book was printed in the United States of America.

Rev. date: 05/06/2014

To order additional copies of this book, contact:
Xlibris LLC
1-888-795-4274
www.Xlibris.com
Orders@Xlibris.com
619026

Contents

Foreword 13

Acknowledgements 17

Introduction 23

Agriculture 25

Industrial Asset Management Through Out The
World—Produces Clean Enviornment 26

Tribology 27

Silt 30

"For Engineers"—Cleanliness Requirements for
Journal Bearing Lubrication 34

Cleanliness chart 37

How Contaminated Oil Destroys Equipment 39

How Contamination Destroys Engines 42

Depth type filtration unit 50

Clean gas (petrol) at fuel pumping stations 51

中文 • Español • Tagalog • TIẾNG VIỆT: www.EnergyUpgradeCA.org/credit

A Message from the California Public Utilities Commission

Look for a Climate Credit from the State of California on Your April Utility Bill

This month your electricity bill will include a credit identified as the "California Climate Credit." Twice a year, in April and October, your household and millions of others throughout the state will receive this credit on your electricity bills.

The Climate Credit is a payment to Californians from a program designed to fight climate change by limiting the amount of greenhouse gas pollution our largest industries put into the atmosphere.

This program is one of many developed as a result of landmark legislation called the Global Warming Solutions Act of 2006, which puts California at the forefront of efforts to battle climate change. Other programs under this law increase clean, renewable forms of electricity, promote increased energy efficiency in homes and businesses, and require cleaner fuels, and cleaner, more efficient cars and trucks.

Together, these programs will aid in reducing greenhouse gas emissions that trap heat in the atmosphere – helping to clean the air and protect our food, water, and public health, as well as the beauty of our state.

The Climate Credit is designed to help you join with California in its efforts to fight climate change and clean the air. You can use the savings on your electricity bills however you choose, but you can save even more money by investing the bill savings from your Climate Credit in energy-saving home upgrades, including more efficient lights and appliances. You can find more information and receive rebates for these and many other energy efficient choices for your home at www.EnergyUpgradeCA.org/credit.

California's greenhouse gas reduction programs provide a range of powerful solutions to help slow climate change, one of the greatest challenges facing society. By gradually reducing emissions each year and moving to cleaner forms of energy, we are taking an important step to preserve the health and prosperity of our state for generations to come.

The CPUC regulates privately owned electric companies and serves the public interest by protecting consumers and ensuring the provision of safe, reliable utility service and infrastructure at reasonable rates, with a commitment to environmental enhancement and a healthy California economy. For more information about our work contact us at: news@cpuc.ca.gov, 800-253-0500, or visit www.cpuc.ca.gov.

Company Profile

The Universal Oil Filtration System 1000 series has been designed after conducting a thorough research and development in the United States of America.

Before the designing phase an in-depth survey was conducted across the various industries, such as, Power & Energy, Iron & Steel & car manufacturing, to understand the working systems. We designed our equipment in the simplest and most versatile manner to suit the diversified needs of various industries.

Universal Oil Corporation (UOC), USA, has been servicing Indian Industries in India for past 14 years. UOC has been associated with large industrial houses like Tatas, Reliance Group, LNJ Bhilwara Group, etc., in **filtration related projects.**

UOC takes pride in introducing itself as the best filtration system provider in whole of Asia. UOC believes in providing best quality systems to its customers, and not just making lofty promises. A list of highly satisfied customers is available.

FOR EMERGING ECONOMIES OF THIRD WORLD COUNTRIES, TO GAIN FROM THIS PRESENTATION, THE MINDSET OF THE PEOPLE HAS TO CHANGE FROM

"WHAT IS IN IT FOR ME"

TO

"WHAT I CAN DO FOR THE COUNTRY AND

OUR PEOPLE"

High officials and engineers have to have open minds to embrace new concepts, which could be very beneficial.

Foreword

Pollution is the ever-growing problem that occurs in all nations across the world.

The information provided here is an attempt to educate the private and public sectors with regard to management of pollution caused by petroleum products.

Cars, trucks, ships etc., use petroleum products for our comfort and ease of transportation. Pollution has to be controlled for the sake of our health.

In diesel fuel, petrol (gas in United States), diesel oil, mobile oil, etc., 1 to 15 micron contamination is very high.

Contamination of 1 to 15 microns per one litre of new oil is approximately 140,000,000 (one hundred forty million) particles. Human hair is about 43 microns.

This should be true for fuels also.

I am surmising that 40% of these particles do not burn inside engines and are spewed out into the air.

Fine filtration of these fuels will produce about 90% pure fuel. This should give better mileage and very little pollution in the atmosphere. Because of this, the cost of these fuels may go up slightly.

Most of the research is & has been on control of sulfur emission, etc., but very little on fine particulates.

Union Of Concerned Scientists in Google Custom Search, in Article Of Particulate Matter [PM], states: **that fine particles less than 1/10th diameter of human hair** pose the most serious threat to human health, as they can penetrate deep into the lungs.

Fine Particulate Matter [PM] pollution is PRIMARY. SECONDARY pollution is from Hydrocarbons, Nitrogen Oxides, and Sulfur Dioxides.

Diesel exhaust is a major contributor to PM.

In 1952, 4000 deaths occurred in London due to severe smog, called black smog.

On 16th March 2014, French Government banned HALF Of ALL THE CARS ON THE ROAD to control pollution. : Ref. Google News.

From Guardian News 19th March 2013: Air pollution, especially from diesel engines, is a neglected hidden killer, and children and very elderly are especially at risk, says Dr. Ian Mudway, a lecturer in respiratory toxicology with environmental research group, at Kings College, London University.

Acknowledgements

I am grateful to my wife Dr. Jagdish Singh [MD], for her support in writing this book as I have a special knowledge of pollutants in oils, and how they can be controlled. A friend of mine in Manchester, Mr. Manmohan Singh Arora, encouraged me a lot, and so did Mr. Deepak Gandhi.

I am very thankful to my brother, Mr. Kulbir Singh, who read and helped edit this book. Also, I acknowledge the contributions of various researchers and engineers from whose works and researches I have taken their input to provide authenticity to the written information in this book.

Causes

Coal

Coal fires, used to heat individual buildings or in a power-producing plant, can emit significant clouds of smoke that contributes to smog. Air pollution from this source has been reported in England since the Middle Ages.*[3]* London, in particular, was notorious up through the mid-20th century for its coal-caused smogs, which were nicknamed 'pea-soupers.' Air pollution of this type is still a problem in areas that generate significant smoke from burning coal, as witnessed by the 2013 autumnal smog in Harbin, China, which closed roads, schools, and the airport.

Transportation emissions

Traffic emissions—such as from trucks, buses, and automobiles—also contribute.[4] Airborne by-products from vehicle exhaust systems cause air pollution and are a major ingredient in the creation of smog in some large cities.[5][6][7][8]

Photochemical smog

Photochemical smog was first described in the 1950s. It is the chemical reaction of sunlight, nitrogen oxides and volatile organic compounds in the atmosphere, which leaves airborne particles and ground-level ozone. [14] This noxious mixture of air pollutants can include the following:

- *Aldehydes*
- Nitrogen oxides, such as nitrogen dioxide
- Peroxyacyl nitrates
- Tropospheric ozone
- *Volatile organic compounds*

characteristic coloration for smog in California in the beige cloud bank behind the Golden Gate Bridge. The brown coloration is due to the NOx in the photochemical smog.

Natural causes

An erupting volcano can also emit high levels of sulphur dioxide along with a large quantity of particulate matter; two key components to the creation of smog. However, the smog created as a result of a volcanic eruption is often known as vog to distinguish it as a natural occurrence.

Health effects

Highland Park Optimist Club wearing smog-gas masks at banquet, Los Angeles, circa 1954

Introduction

In late 1960's California started implementing clean air enforcement. All the diesel vehicles in California or entering California had to have filtered diesel fuel.

This was strictly monitored and enforced. In Southern California, especially in Los Angeles and surrounding areas. There were days that you could not see mountains 3 miles away. The air smelled and irritated your eyes & throat.

In late 1980's the air became cleaner: you could see the mountains and the eyes and throat did not hurt.

Even though the air is much cleaner, there are days when there is smog in Los Angeles.

Fuel has to be filtered and 1 to 15 micron contamination particles have to be decreased by further filtration with 1 micron filters. This will further decrease smog in Los Angeles.

The fuel has about 30% to 40% contaminants that do not burn, and are thrown into the air, causing pollution. This pollution causes various types of sicknesses.

In economically growing countries like China, India, Thailand, etc., pollution spewed out by petroleum products is causing ailments like Asthma, Eye Irritation, Sore Throat, Cancer, and other dangerous diseases.

All these ailments could be lessened considerably with a clean environment.

Agriculture

Various types of industrial oils are dumped on the ground in these growing economies causing considerable damage to water and air & resulting in less fertile ground. This damages the quality of produce, which, in turn, causes health related problems.

Industrial Asset Management Through Out The World— Produces Clean Enviornment

Would you like to have a relatively maintenance free production in Plants where oil is used extensively in equipment & for the production of different products?

Millions of dollars are spent in procuring different equipment in various industries.

It is a pity that engineers are not aware of how to prolong the life of machinery and oil used in it. Power Plant turbines, hydraulic systems, car engines, equipment having gear oil are examples.

Tribology

CONTROL OF 1 to 15 MICRON Contaminants

TRIBOLOGY IS A SCIENCE OF FRICTION CONTROL to ENHANCE EQUIPMENT LIFE.

Tribology and Its Essential Progress in the Successful Operation of Machines

The subject of this writing will explain a term that is relatively new based on an element in Greek that is used in modern engineering and physics:

Tribology. This Greek **tribo**—element means "friction, rub, grind" or "to wear away".

Lubrication is central to machine performance, but it's only one part of the story. More and more, the bigger picture of machine health has been ignored.

- Tribology combines issues of **lubrication**, **friction**, and **wear** into a complex framework for designing, maintaining, and trouble-shooting the whole machine world.

- Tribology is already providing data that could be used to produce transmission fluids that give automobile drivers better fuel economy and a smoother ride.

- The most visionary tribology advocates and practitioners tend to view their field as the cure for much of what ails industry and even entire economies.

- *Oxford English Dictionary* defines tribology as, "The branch of science and technology concerned with interacting surfaces in relative motion and with associated matters. (like friction, wear, lubrication, and the design of bearings).

Nomenclature:

NAS—National Aerospace Standard.

Close Tolerance: Extremely close gaps between two moving parts.

Lapping compound: Like Emory Paper Rubbing.

Journal Bearing: Like a rod having bearings at both ends.

Silt

Particles of 1-2 microns are not considered in **NAS level** determination. However, they are just as dangerous as 5-15 micron particles.

UOC filtration removes large numbers of 1-2 micron particles, known as **silt**, which cause **considerable wear** to close tolerance parts and cause **internal leakages**, thereby **lowering system efficiency**.

The oil in the system travels at **very high pressure and speed and the silt causes considerable wear to the valves and seals**.

These silt particles act like bullets, causing **Lapping Compound like action**.

By controlling silt, **friction** in machinery is considerably reduced, thus giving longer life and trouble free production.

- 85% of problems in equipment is due to highly contaminated oil with silt.

- **Increased equipment life means millions of dollars saved.**

Silt in New oil [**1-2 micron** particles], is approximately **13,000,000/100 ml.**

In used oil, silt could be as much as **3 to 7 times more**.

Therefore, used oil, even brought to NAS "4", would have a wear rate at least **10 times more than New oil [just like dirty oil]**, unless 1-2 micron particles are **brought down to the level of New oil**.

Transmissions are just one place where tribology makes a difference in the automotive industry.

Other items on the agenda include controlling brake noise and wear, reducing internal friction in engines, increasing the productivity, part quality, and energy efficiency of production machinery.

What Is Silt Lock?

Silt lock is when micron-sized particles (silt) become lodged between the hydraulic valve spool and the bore. Silt particles migrate into the clearances between the spool and bore, increasing friction when the valve is actuated when more and more of these silt particles become lodged in the clearances. It eventually results in silt lock.

Silt lock stops production, increases valve maintenance costs and slows production due to sluggish response. Setting up a lubrication program to control contamination will prevent this from happening.

The presence of varnish on valve spools and bores tightens the interference fit (annular clearance), reducing the particle size affecting contaminant lock.

The varnish also has adherent properties that stick the particles to the silt lands.

The longer a valve holds pressure without actuation, the longer the available time for the valve to silt up (and sludge up).

Most stiction-related valve failures occur just after a long dwell time. Large amounts of silt-sized particles in the 1—to 5—micron range have a tendency to grow dramatically in population as oils age.

These clearance-sized particles increase the propensity of contaminant lock.

Water has a tendency to preferentially coat particles. Two such particles in contact will cling together (like wet sand), aggravating the silt-lock risk considerably.

"For Engineers"

Cleanliness Requirements for Journal Bearing Lubrication

Refrences:

Kelly Collins, Pall Corporation *John Duchowski,* Pall Corporation

The study of journal bearing

The minimum film thickness is found at the closest point of contact between the journal and the sleeve.

Theoretical analysis reveals that for these journal-bearing sizes under normal operating conditions, a 1-20 microns thick hydrodynamic film forms to separate the journal from the pads, since film thickness is dependent on fluid viscosity.

As the fluid temperature rises, the film thickness may be reduced by as much as 20% for some lubricants.

Theoretical analysis shows the minimum film thickness is between 1-20 microns; however, empirical results reveal that even though dimensional clearances within a journal bearing may differ, and the load and rotational speed may vary, the **actual film thickness is in the order of approximately 10 microns**.

Abrasive wear occurs when **clearance-sized particles come between two surfaces** that are **under load**, such as in the journal and the bearing.

These tests were conducted over a 20-hour period with interruptions for wear measurements, at five hour intervals.

It can be seen that **more than a ten-fold increase in bearing wear** results from contaminated oil.

The filtration requirement is most critical at the commissioning of turbines, compressors or other equipment and when the equipment **is rotating at low rpm**. It is at these times that the **hydrodynamic film is the thinnest.**

Unfortunately, some equipment designers tend to consider the overall capital costs rather than the technical requirements of the system when choosing filtration systems. Field experience has proven that this approach can result in much higher operating costs when the costs for repairs, maintenance, parts replacement and lost production are considered.

Given that the film thickness that exists under normal operating conditions is **approximately 10 microns**, and since **the film thickness is further reduced during startup** and at low viscosity condition or with lower viscosity fluids, it is recommended that the filter employed exhibit a high removal efficiency of particles **down** to 5 microns **or finer**.

DESCRIPTION OF UOC FILTERS

UOC Filters are specially designed to reduce the friction in the oil by filtering out all the contaminants from the oil. These filters are called Depth Type Filters. Beta ratio of these filters is 99.99. Most of the oil systems found in the industries are at NAS 12+ level which in itself has more than 1 million particles of 5 to 15 microns. Even new oil available in the market is at NAS 10-12 level. UOC filters not only filter out these and larger particles but they also filter out particles of 1 and 2 micron sizes which are not considered for NAS value calculations. Given below is the contamination chart:

New Oil

ISO Code	9/6	10/7	12/9	13/10	14/11	16/13	21/18
National Aerospace Standards (NAS)	0	1	3	4	5	7	12
Size Range (Micron)	Values based per 100 ml						
5-15	250	500	2000	4000	8000	32000	1024000
15-25	22	88	356	712	1425	5700	182400
25-50	4	16	63	126	253	1012	33400
50-100	1	3	11	22	45	180	5760
Over 100	0	1	2	4	8	32	1024

SILT

System efficiency:

The oil in the system travels at very high pressure and speed, and the silt causes Particles of 1-2 microns are not considered in NAS level determination. However, they are just as dangerous as 5-15 micron particles. UOC filtration removes large numbers of 1-2 micron particles, known as silt, which cause considerable wear to close tolerance parts and cause internal leakages, thereby lowering considerable wear to the valve seals. These silt particles act like bullets, causing Lapping Compound like action. (Emory Paper Action)

How Contaminated Oil Destroys Equipment

Solids suspended in oil are like grinding paste.

They scour and gouge surfaces; block oil passages and makes the oil more viscous. The longer the oil is left dirty the faster the rate of failure.

Many **original equipment manufacturers** have accepted the indisputable evidence from numerous field and laboratory trials that **oil cleanliness has a major effect on wear within their equipment.**

Some of them are now specifying how clean the oil used in their equipment should be if warranty claims are to be honoured.

For example Caterpillar specifies new oil to have a particle count of ISO 16/13 or NAS 7.

If new oil is above this level of contamination they will not warranty their equipment.

When new oil from a leading international oil manufacturer was tested before putting it into new Caterpillar equipment, the solid particle contamination was found to be 17/14. This was new oil from a never previously opened container.

In this case the new oil had to be further filtered to bring it to below the required specification.

If you want extremely low wear rates and long equipment life the evidence indicates that oil needs to be filtered down to sub 5 micron size and preferably down to one micron size.

For **expensive hydraulic and oil lubricated equipment filtration cost is easily and quickly recuperated by the large gain in equipment working life and reliability.**

How Contamination Destroys Engines

HOW CONTAMINATION DESTROYS ENGINES

New oil is NAS 12 Fig. (A) Directly below, the black patch is figure (C) is oil tested after running engine on test stand for 45 minutes. This oil has become unsafe. It should be filtered with 1 micron Depth Filter or replaced with new oil. Figure (D) is cleaner than new oil in Figure (A), after filtration of Figure (C) oil, On the extreme upper right figure (E), 1 micron depth filter attached to engine while running on test stand, resulting in very clean oil.

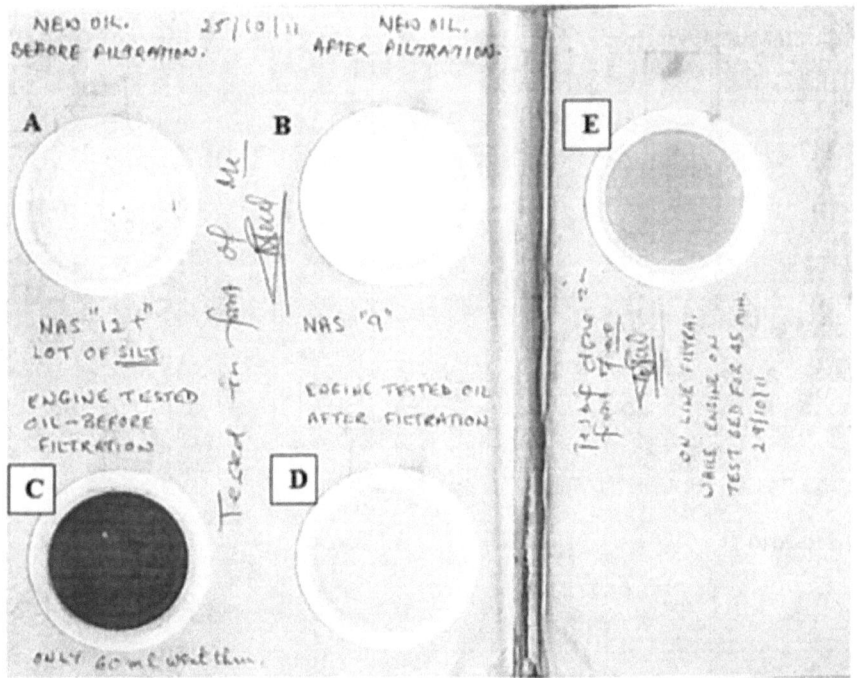

NAS 12+

LUBE A

SAMPLE POINT – 615

MMABOVE TANK

BOTTOM

320 VISCOSITY

40 MR. PASSED

THOUGH MEMBRANE

26-10-2010

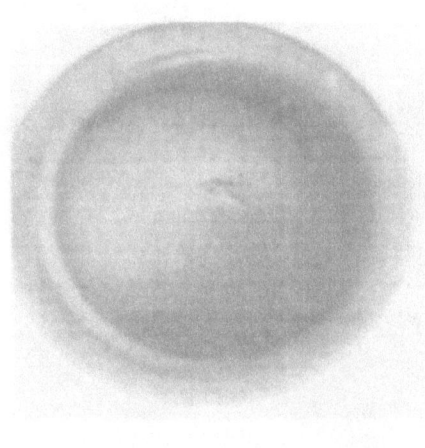

NAS 12 +

LUBE B

TAK CAPACITY – 2500 LTRS

320 VISCOSITY

40 ML PASSED THOUGH MEMBRANE

26-10-2010

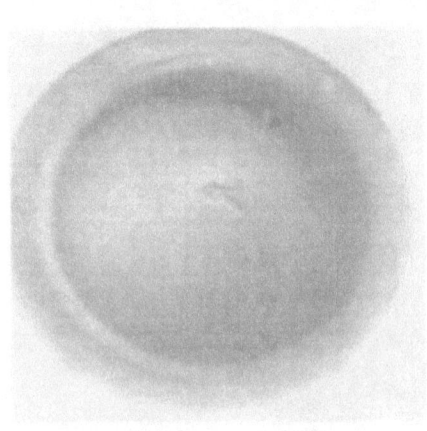

Lube A – Mill Drive Gear Box

Lube B – RES & Cold Shear

Lube "A" oil sample was taken from Pump Suction Chamber [615 mm up from bottom of tank]. This oil is After Filtration from an external filtration unit. Millipore test was conducted in front of Mr. V Kudva. Only 40 ml of fluid passed through the membrane. The rest of the fluid [60 mm] was discarded.

Lube "B" oil sample was tested. Again only 40 ml of fluid passed through the membrane. Sample was taken from the sampling point, near the drain point. To this system no external filtration is attached.

Test Results :
Tremendous amount of Silt
Shiny metal particles
Oil is NAS 12+

Comments—Oil is not fit for use, and has already started damaging the internals of both the systems.

Solution—Oil should be drained from both the systems and stored in drums. After couple of weeks the contamination will settle at the bottom of drums, and the oil then syphoned off 10" from drum bottoms, and oil put into fresh drums. This oil can then be used as fresh oil for topping up, etc.

With silt control—no wear at all (no scoring at all—shaft is smooth)

Figure 2. **Silt** Control

Oil is NAS 4-5

This picture was magnified

without silt control—wear on spool (scoring on the spool)

Figure 1. **No Silt** Control

Oil is NAS 4-5

Pictures Provided By Tata Steel After 5 Years Of once A Month Filtration.

Universal Oil Company has been doing
Turbine and Hydraulic oil filtration for last 14 years.

Universal Oil Company case study on silt control. [2006]

Ref: SMMM/285/13
Date: 4/10/13

To whom it may Concern

TATA STEEL FINDINGS OF BENEFITS OF FILTRATION, BY USING **UNIVERSAL OIL COMPANY** [UOC] **FILTRATION UNIT, THAT ALSO CONTROLS SILT** [1—5 micron contamination].

Using Universal Oil Company [UOC] Filtration Unit, we have found that system performance of Hydraulic Power Packs and Controls, and of Captive Power Plants, Turbine and Turbine Controls, have **increased**.

We also found:

a] increased life of components immersed in oil.
b] Increased life of built-in filters in our systems.
c] increased life of oil, for an extended period due to control of silt. With proper filtration, all oil properties remain the same.
d] Wear and tear of components immersed in oil, slowed down
e] No more leakages from joints, seals etc., and provided damaged parts are replaced.
f] Less maintenance manpower utilized.
g] Maximum efficiency for production is ensured.

NOTE: We found **no adverse effects** in all our systems due to control of silt and maintaining NAS level below 5

The above benefits give **more profits** to the company.

Sunil Mishra
4/10/13

Sunil Kumar Mishra
(Chief, Steel Making Mechanical Maintenance)

Chief Mech. Maint. (Steel Making)

TATA STEEL LIMITED
Jamshedpur 831 001 India
Registered Office Bombay House 24 Homi Mody Street Mumbai 400 001

To whomsoever it may concern:

Universal oil Company has performed filtration of our Press power pack and has brought the system oil to NAS 7 from NAS 14, in an environment that is extremely dirty, where fine dust from graphite is present in the air. Also simultaneously controlled silt [1 to 2 micron contamination particles], which is very harmful to all the components immersed in oil.

We have now purchased the filtration unit from UOC and attached it permanently to the power pack. Now the system oil is maintained to NAS 6.

We are satisfied with the performance of UOC filtration unit.

Sanjay Dubey

ZION STEEL LIMITED

Vill: Gobira, P. O.: Kuarmunda, Dist.: Sundargarh-770039, Orissa.
E.mail-steelzion@gmail.com

TO WHOM IT MAY CONCERN

This to certify that M/S Universal oil filtration has been extremely beneficial in our Rolling mill-2

The life of the equipment and availability improved by 20%, from last six months our equipment running smoothly. The equipment breakdown reduced and quality has also been improved.

The details of power packs given below:

Function

- Pusher power packs -   For pushing of billets inside reheating furnace
- Babbit metal bearing -   The force lubrication of babbit metal bearing
- Tilting table -   The operation of tilting table movement
- Cooling bed brake slide apron -   The up down movement of cooling bed aprons
- Twin Channel -   The openings and closing of twin channel flaps

We satisfy with the results
MANOJ KUMAR TIWARI
AVP_(MILLS)

05/02/14

Regd. Office: 14, Netaji Subhash Road, 2nd Floor, Kolkata-700001
Ph. No.: 033-22102105, Fax: 033-22102105

DEPTH TYPE FILTRATION UNIT HAVING 1 MICRON PAPER FILTER USED FOR CONTROLLING CONTAMINANTS OF 1 MICRON & UP.

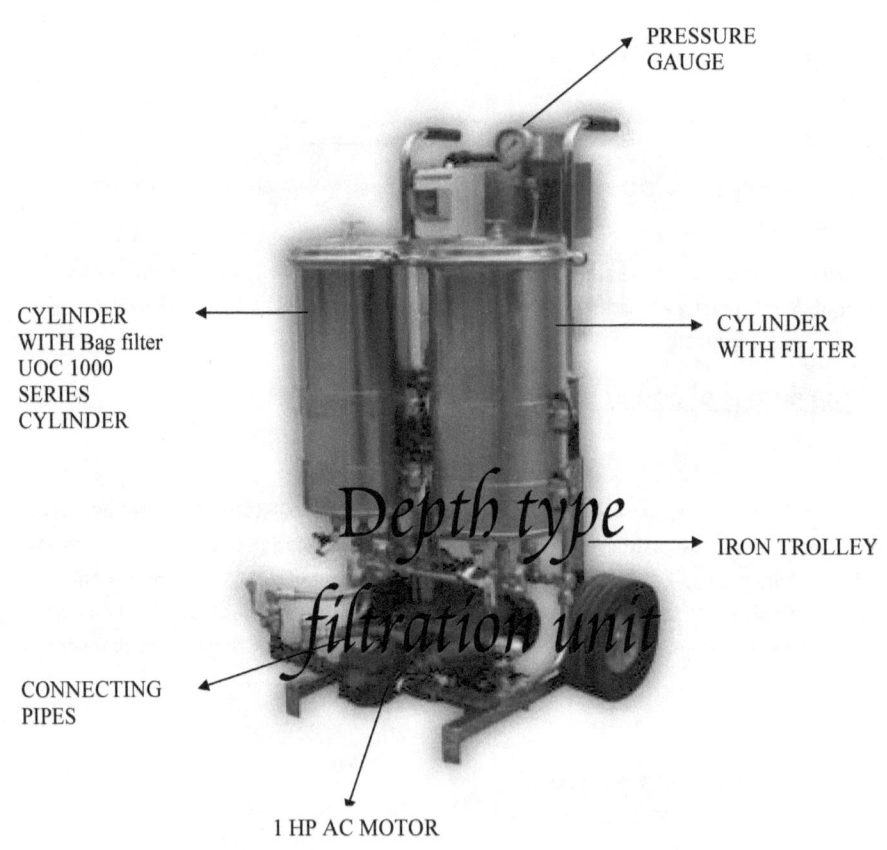

REQUIREMENT FOR GAS PUMPS (PETROL PUMPS), TO CONVERT FOR DISPENCING SUPER CLEAN FUELS TO VEHICLES. FOR (PETROL PUMPS) GAS STATIONS TO CONVERT, REQUIRE 2 TANKS OF 1.5 AND 2.0 CUBIC METERS.

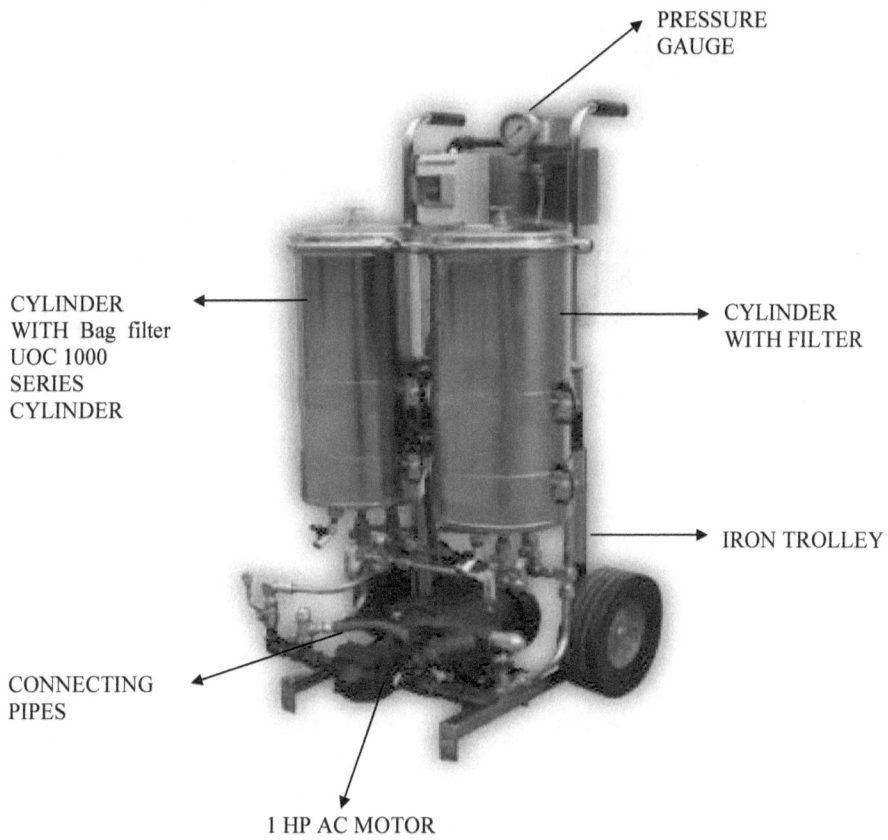

UNIVERSAL OIL COMPANY (UOC) IS AVAILABLE FOR CONSULTATION

The two tanks should be connected to each other, and 2.0 cubic tank (A) should be connected to the fuel tank underground.

The fuel is pumped from the underground tank into tank A, where the fuel will be filtered and pumped into smaller tank (B).

Tank B will be connected to petrol (gas or fuel) pump, from where the fuel will be pumped into vehicles.

One micron depth filter should be used for filtration.

www.ingramcontent.com/pod-product-compliance
Lightning Source LLC
Chambersburg PA
CBHW021045180526
45163CB00005B/2287